ANCIENT TECHNOLOGY

ANCIENT COMPUTING

Ancient Technology

ANCIENT COMPUTING

FROM COUNTING TO CALENDARS

by Michael Woods
and
Mary B. Woods

RP Runestone Press • Minneapolis
A division of Lerner Publishing Group

Dedicated to Betty Mulhearn Morgan Kearney

Series designer: Zachary Marell
Series editors: Joelle E. Riley and Dina Drits
Line editor: Margaret J. Goldstein

Runestone Press
A division of Lerner Publishing Group
241 First Avenue North
Minneapolis, MN 55401 U.S.A.

Website address: www.lernerbooks.com

LIBRARY OF CONGRESS CATALOGING-IN-PUBLICATION DATA

Woods, Michael, 1946–
Ancient computing : from counting to calendars / by Michael Woods and Mary B. Woods.
p. cm. — (Ancient technology)
Includes bibliographical references and index.
Summary: Discusses the methods of computation developed in various civilizations around the world, from prehistoric times up until the end of the Roman Empire.
ISBN 0-8225-2997-1
1. Mathematics, Ancient—Juvenile literature. [1. Mathematics, Ancient. 2. Mathematics—History.] I. Woods, Michael, 1946–. II. Title. III. Series.
QA22.W66 2000
510'.93-dc21 99-39810

Manufactured in the United States of America
1 2 3 4 5 6 - AM - 05 04 03 02 01 00

TABLE OF CONTENTS

Introduction / 6

Chapter 1, The Stone Age / 11

Chapter 2, Ancient Middle East / 17

Chapter 3, Ancient Egypt / 27

Chapter 4, Ancient India and China / 41

Chapter 5, Ancient Americas / 53

Chapter 6, Ancient Greece / 61

Chapter 7, Ancient Rome / 77

Glossary / 84

Selected Bibliography / 85

Index / 86

What do you think of when you hear the word *technology?* You probably think of something totally new. You might think of supercomputers, powerful microscopes, and other scientific tools. But technology doesn't mean just brand-new machines and discoveries. Technology is as old as human civilization.

The word *civilization* comes from the Latin *civitas,* meaning city. However, people must do more than build cities to be civilized. Civilizations also have governments, religions, social classes, distinct occupations, and methods for keeping records.

Technology is the use of knowledge, inventions, and discoveries to make life better. The word *technology* comes from two Greek words. One, *tekhne,* means "art" or "craft." The other, *logos,* means "word" or "speech." Ancient Greeks used the word *technology* to mean a discussion of arts and crafts. Today, *technology* refers to an art or craft itself.

People use many kinds of technology. Medicine is one kind of technology. Construction and agriculture are also kinds of technology. These technologies and many others make life easier, safer, and happier. This book looks at a form of technology used in almost every area of daily life and every field of science. That technology is computing.

What Is Computing?

When people hear the word *computing,* they often think about using computers. But *computing* has another meaning.

Computing involves using numbers to count, solve problems, and gather information. Computing also involves manipulating numbers by adding, subtracting, multiplying, and dividing them. Computing can be as simple as 1 + 1 = 2. Computing can also be difficult, involving weeks of time and requiring the help of supercomputers.

Computing involves mathematics—the science of numbers. Math has many branches and many practical applications. It is used in almost every area of science, medicine, business, construction, manufacturing, and mapmaking.

Ancient Roots

You've probably heard people remark, "There's nothing new under the sun!" That's an exaggeration, of course. Yet there's much truth in the saying when we're talking about computing. Computing probably began shortly after the first humans appeared on earth. Stone Age people performed mathematics by counting, probably on the fingers of one hand. They kept track of numbers by cutting notches into sticks and tying knots in ropes. This first computing technology was primitive, yet it was effective, easy to learn, and accurate.

The word *ancient*, as used in this book, refers to the period from the emergence of the first humans on earth to the fall of the Western Roman Empire in A.D. 476. The first human beings lived about 2.5 million years ago.

CIVILIZATIONS OF THE Ancient World (through A.D. 476)

EUROPE
ASIA
AFRICA
Indian
Ocean
6000 B.C.
534 B.C.
Middle East
3100 B.C.
30 B.C.
Egypt
1766 B.C.
China
1200 B.C.
Mesoamerica
800 B.C.
146 B.C.
Greece
509 B.C.
A.D. 476
Rome
320 B.C.
India
Stone Age civilizations have flourished in most parts of the world. These cultures began and ended at different times in different regions.

Technology is one trait that scientists use to distinguish humans from their prehuman ancestors. Experts say that our ancestors became fully human when they started using technology such as counting.

Ancient people discovered some computing methods by trial and error. Sometimes people copied and improved on computing technology used in other countries. The ancient Greeks, for instance, learned some of their computing knowledge from the Egyptians and the Babylonians. The Romans learned from the Greeks. Each civilization added improvements. Gradually, computing knowledge spread throughout the world. Mathematics became a universal language.

A Lot with a Little

Ancient people did not have personal computers, pocket calculators, or even adding machines. But they performed very effective computations. Most of the major branches of mathematics originated in ancient times. Ancient engineers and builders used math to design roads, buildings, machines, and weapons.

Ancient computing has stood the test of time. We measure angles in degrees, minutes, and seconds thanks to the ancient Babylonians. We divide our day into 24 hours, just as the ancient Egyptians did. Our number symbols were created in ancient India. We sometimes use Roman numerals, symbols for numbers developed more than two thousand years ago.

Ancient people used math for fun, too, and developed number games, tricks, and puzzles. Read on and discover many wonders about the computing knowledge we inherited from people who lived thousands of years ago.

chapter 1 one

THE STONE AGE

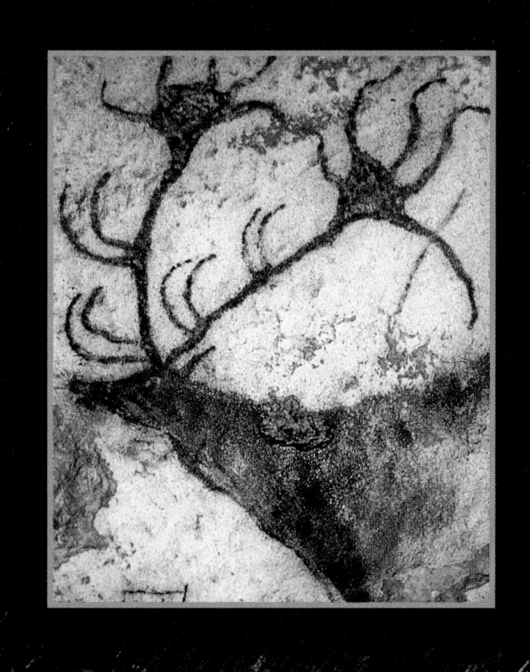

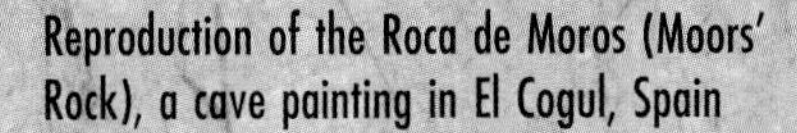

Reproduction of the Roca de Moros (Moors' Rock), a cave painting in El Cogul, Spain

The first period in human society is called the Stone Age. The first humans were hunters and gatherers: people who obtained food by hunting, fishing, and gathering wild plants. They lived in small groups that moved from one area to another. When the food in one area was used up, the group moved on. People lived this type of nomadic existence until about 10,000 B.C.

Stone Age people probably knew the importance of quantities, or amounts. They knew that two antelopes meant more food than one. A pack of wolves was more dangerous than a lone wolf. A bunch of berries was more valuable than one berry. But did Stone Age people understand the ideas behind numbers?

Three pencils, three cars, and three stars in the sky all have something in common: "threeness." Ten mastodons, ten birds, and ten trees all share the trait of being ten in number. By understanding this connection, ancient people could create symbols used for counting. That is, any group of three objects could be described with a certain symbol. A different symbol could stand for a collection of two, four, and so on.

Finger Symbols

We can only guess about when Stone Age humans developed basic systems for counting. They probably used fingers to represent numbers, just as young children do when they learn to count. One finger was probably the universal symbol for 1, two fingers for 2, three for 3, and so on. For Stone Age hunters, four extended fingers might have stood for four woolly mammoths hiding just out of sight.

It's no surprise that our modern numbering system is based on 10—the number of human fingers. In fact, the word *digit*, meaning a single numeral, also refers to a finger or toe.

Sticks and Cords

Ancient people kept track of numbers in another way—with notches cut into sticks. Archaeologists, scientists who study the remains of past cultures, have found Stone Age tally sticks—sticks and bones with rows of neatly cut notches. *Tally* means to add or to keep track of amounts. Ancient hunter-gatherers may have used tally sticks to keep track of amounts of food collected by each person in a group. By keeping a tally, each person could claim his or her fair share.

Stone Age people also recorded numbers by tying knots in cords. "Talking cords" were lengths of vine, rope, or leather. Each knot in the cord represented a number. Imagine a shepherd counting sheep by tying a knot each time one animal passed by. Maybe the shepherd devised a system of counting by twos or threes, tying a knot for every two or three sheep that passed by. Talking cords were used for millions of years and are still used in some countries.

Measuring with Body Parts

The first measuring sticks were parts of the human body. For instance, ancient people would use their feet to pace

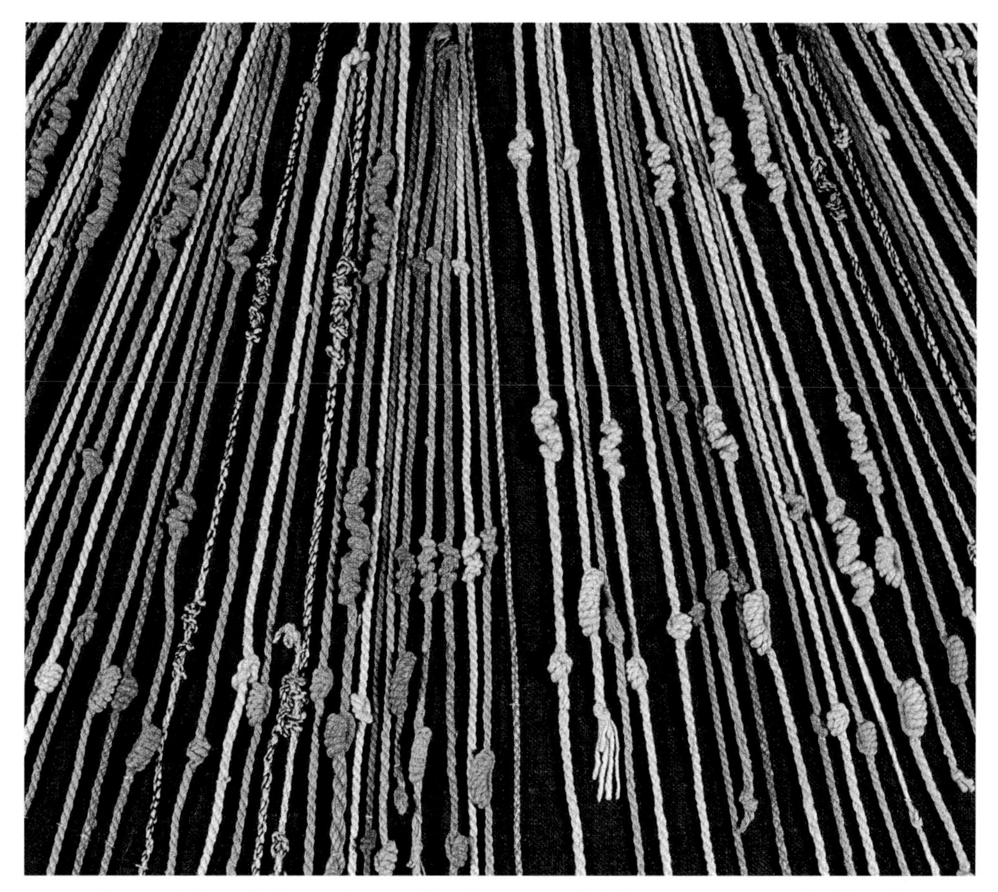

Talking cords from the Inca civilization of South America, which became an empire in the fifteenth century A.D. Talking cords are still used in some cultures.

off distances, one foot placed directly in front of the other. For thousands of years, the *foot*—equal to 12 inches in modern times—was not a fixed length. It varied by as much as several inches, depending on the size of the human foot doing the measurement.

One of the most widely used ancient units of measurement, the *cubit*, was the distance from a man's elbow to the end of his middle finger. One *inch* was at first the width of a man's thumb, later the length of an index finger from the tip to the first joint. The *hand* was the width of a man's hand—about four inches. (The height of horses is still measured in hands.)

Body-part measurements were not uniform. They varied a lot from person to person. But they did offer a big benefit—ancient people always had a ruler handy.

chapter two

ANCIENT MIDDLE EAST

Relief of merchants, made in the seventh century B.C. in the city of Carchemish, Turkey

Around 3500 B.C., people in the Middle East began to abandon the hunter-gatherer lifestyle. They became more settled, building houses, farms, and villages in a rich growing region known as "the Fertile Crescent." The area was home to many ancient civilizations, including the Mesopotamians, Sumerians, Babylonians, Hittites, Assyrians, and Phoenicians.

One technology often leads to the development of another. Because people in the ancient Middle East were farmers, they needed methods for counting crops, measuring land, and keeping track of the growing season. Because they traded goods with other groups of people, they needed scales, uniform systems of measurement, and other forms of computing technology.

One Thing Leads to Another

As people in the Middle East settled into villages and built homes and farms, they soon needed to mark the boundaries of their land. They developed a technology called surveying, using math to measure land features and to determine borders between lots. With surveying techniques, ancient mapmakers could accurately show rivers, hills, and other land features on maps. Surveying was also important in construction. It helped ancient engineers design straight roads, buildings, and bridges.

Writings and other artifacts left by the Sumerians show that people in the Middle East measured land boundaries as early as 1400 B.C. The Sumerians also built their cities according to a plan, which required careful measurement and surveying techniques.

Boundary stone of King Nazimaruttash of Babylonia, made in the thirteenth century B.C.

The First Maps

Maps are charts that show the location of cities, roads, and land features such as mountains and rivers, and the distances between them. Mapmaking requires very precise measurement. Mapmakers must draw maps to scale, meaning that distances on maps are proportional to distances in the real world. One inch on a map, for instance, may equal 10 miles on land.

The ancient Babylonians drew the first maps, inscribed on clay tablets, about 2300 B.C. Many of these maps were official records of landownership and showed the size of farmers' fields. Other maps were guides for people on long journeys. One ancient Babylonian map, drawn around 600 B.C., showed the entire world—or at least what the Babylonians thought was the entire world. It was a very small place, with Babylon in the center, the Persian Gulf off to one side, and a few other countries, such as modern-day Armenia. All the land was surrounded by a huge ocean.

The First Salespeople

The Fertile Crescent contained excellent farmland, and farmers in the ancient Middle East produced more food than they needed. Thus they were able to sell the surplus.

Babylon was the first true commercial center. From there, merchants traded food and other products with people in many other cities. To charge and pay the same amount for identical amounts of grain, dried fish, cloth, brick, gold, and other products, merchants needed standard units of money, length, and weight.

The cubit, the distance from a man's elbow to the tip of his middle finger, was a widely used unit of length in the

Sumerian tablet with pictogram records of malt and beer. The pictogram for beer is a conical jug.

ancient world. The Mesopotamians divided the cubit into smaller units. One cubit contained two feet, and one foot contained three hands—each of which was the distance from index finger to little finger. A fingerwidth was equal to about one inch in our modern system of measurement.

Computing Weight

A scale is a device used for weighing objects. Archaeologists are not sure whether the first scale was invented in ancient Babylonia or in ancient Egypt. Both civilizations used scales, perhaps as early as 5000 B.C.

Ancient scales were beam scales made from a stick or rod that balanced upon a center support. A pan hung from each end of the beam. When an object in one pan (a piece of gold, perhaps) was heavier than an object in the other, one pan hung lower. When the objects were equal in weight, the pans balanced.

The first beam scales simply compared the weight of two different objects; they didn't measure an object's weight based on standard units. The Babylonians eventually developed the world's first weight standards—units of measurement that were the same from place to place and never changed.

The Babylonian standards were smooth stones, ground

Four-thousand-year-old Babylonian hematite weights, ranging from a mina (500 grams) to a shekel (25 grams)

and polished to ensure that each weighed the same. Merchants placed one or more stones on one pan of a beam scale and placed objects to be bought or sold in the other. With standardized weights, business transactions were made more accurate.

The beam scales used in Babylon may seem primitive. But modern scientists still use beam scales to measure tiny amounts of material. Some modern beam scales are accurate to about 1 millionth of a gram—or about 400 millionths of an ounce. The scales are so accurate that a speck of dust can affect a measurement. Only some kinds of electronic scales are more accurate.

Computing Time

Sundials are devices that measure time by recording the position of the Sun as it moves across the sky. Sundials are very accurate timekeepers, and they are still used in some countries.

The first sundials were made in the ancient Middle East. They were flat pieces of stone or wood, affixed with an upright bar, called a gnomon, that cast a shadow. As the Sun moved across the sky, the gnomon's shadow moved across lines inscribed on the sundial's base, each line representing a certain time of day.

Around 300 B.C., a Babylonian astronomer named Berosus made a sundial with a curved base, sort of like a bowl. Using 12 lines on the base of the sundial, Berosus divided the day into 12 equal intervals. These were the first hours. Berosus's clock was so good that others like it were used for more than a thousand years in countries throughout the Middle East. Our modern system of measuring

24-hour days, with 12 hours in the morning and 12 hours in the afternoon and evening, began with Berosus's 12-hour system.

Counting by 60s

We measure time in units of 60—a minute has 60 seconds, an hour has 60 minutes. Degrees, units of measurement for angles, are also based on the number 60—a circle contains 360 degrees, which is 60 × 6. Where did the use of 60 as a base measurement begin? It began with ancient Mesopotamia's numbering system.

Archaeologists have found clay tablets marked with numerals in the ruins of Babylon and other ancient Middle Eastern cities. Sixty was the basis of the numbering system—symbols on the tablets stood for 1, 10, 60, 600, 3,600, and 36,000.

A Space Monster Is Eating the Sun: Part One

During an eclipse of the Sun, the Moon passes between the Sun and the Earth, blocking the Sun's light. During an eclipse of the Moon, the Earth passes between the Sun and the Moon. Earth's shadow darkens the Moon. To ancient people, eclipses were mysterious and frightening. Some ancient people, such as the Chinese, believed that a space dragon or monster swallowed the Sun or Moon during an eclipse.

Babylonian astronomers wanted to know when eclipses would occur so that they could alert their rulers. For centuries, the astronomers recorded the dates of eclipses and the movements of heavenly bodies. Ancient scientists did not understand why eclipses occurred, but they knew when to expect them. Their work was the start of what we call mathematical astronomy.

A Famous Number

One of the most useful numbers to engineers, physicists, and other scientists is *pi*. Multiplying the diameter of a circle by pi gives you the circle's circumference. The diameter is the distance from one side of the circle to the other, going directly through the middle. The circumference is the distance around the circle.

By studying how the circumference of a circle changes as its diameter changes, the ancient Babylonians and Egyptians discovered pi around 2000 B.C. Babylonian mathematicians figured that pi was equal to 3.125. Egyptians figured the number at 3.160. Modern mathematicians define pi as approximately 3.1416.

Pi is equal to 22/7, but it is impossible to convert this fraction into an exact decimal number. When you divide 22 by 7, the numbers after the decimal point just go on and on forever. Using supercomputers, mathematicians have figured pi's value to more than six billion decimal places—that's 3, then a decimal point followed by six billion numbers.

Pi was one of the greatest discoveries in computing history. Pi works on each and every circle, no matter what the size. Using pi, ancient people could compute the distance around any circular field, building, or other object by measuring the circle's diameter and multiplying by 3.1. With a different equation, pi could be used to find the area enclosed by a circle. Pi has many other uses in areas of math, engineering, and physics.

chapter 3 three

ANCIENT EGYPT

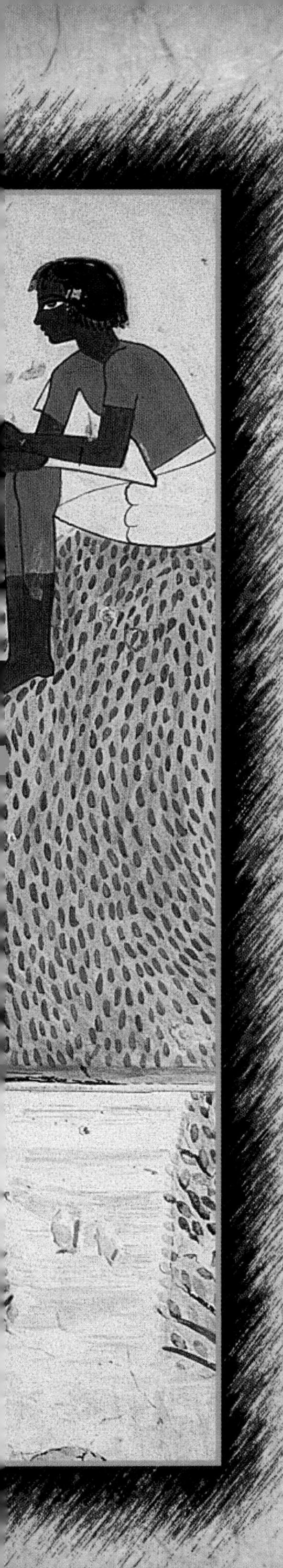

Wall painting of workers measuring and recording the year's crop, from the Tomb of Menna in Thebes, Egypt

Civilization in Egypt developed along the Nile River, the world's longest, in about 7000 B.C. Ancient people built permanent settlements and farms along the river, which provided water for drinking, bathing, and growing crops. As another benefit, the Nile overflowed its banks every year, depositing a layer of muck that fertilized the soil. Gradually, the Egyptians developed one of the ancient world's most famous civilizations, renowned for its giant pyramids, a picture-writing system called hieroglyphics, and other advanced technology.

The ancient Egyptians used computing technology for many projects. They used addition and subtraction to keep track of taxes and business transactions. They used surveying techniques to measure farmers' fields. They used sundials and other types of clocks to measure time. They used engineering techniques to build giant pyramids and temples.

Picture-Numbers

Mention hieroglyphics and most people think of the Egyptian system of picture-writing. But hieroglyphics was also picture-numbering. In the Egyptian system, a single line stood for 1, two lines for 2, three lines for 3, and so on up to 9. An archlike symbol stood for 10, and 100 was represented by a spiral. The number 1,000 was represented by a lotus plant, 10,000 was an index finger, 100,000 was a tadpole or frog, and 1,000,000 was a man sitting with arms upraised.

To write the number 1,109, then, the Egyptians would draw a lotus plant (1,000), a spiral (100), and nine lines (9). One finger, one lotus, and two spirals meant 11,200. A man and a tadpole equaled 1,100,000.

Ancient Textbooks

Imagine this: Thousands of years from now, archaeologists discover the remains of your school classroom. As they dig through the rubble, they find your math textbook and put it into a museum. Scholars around the world study it. Why? Because your textbook contains the only remaining knowledge of mathematics in the twenty-first century.

Sound unrealistic? That's exactly what happened when archaeologists discovered two textbooks used in schools in ancient Egypt. Both books were long scrolls of papyrus, a kind of paper used in ancient times. The books were used to teach scribes—professionals who were trained to read, write, and perform equations in ancient times. One of the scrolls, the Rhind Papyrus, was used in the seventeenth century B.C. The other scroll, the Golenishchev Papyrus, was used in the nineteenth century B.C.

Relief showing numerals, from temple complex at Karnak, Egypt

Ahmes the Moonborn

The Rhind Papyrus is our most important source of information about Egyptian math. It was named for Alexander Henry Rhind, a Scottish archaeologist who discovered the scroll near the Egyptian city of Thebes in A.D. 1858. When unrolled, the scroll is about 18 feet long.

An Egyptian scribe, Ahmes the Moonborn, wrote the papyrus around 1650 B.C. Entitled "How to Obtain Information About All Things Mysterious and Dark," the papyrus explained how to add, subtract, and do other computations with whole numbers and fractions. Most ancient Egyptians were not educated and wouldn't have understood the scroll, which explains why its contents were considered "mysterious and dark." But the equations would be a snap for most modern sixth-grade pupils.

Ahmes also included more advanced math in his textbook, including algebra, a branch of math that deals with quantities expressed in symbols. One simple algebra equation is $6 + x = 7$. What is x? The answer, of course, is 1. Another algebra equation is $45 - x = 40$. The answer: $x = 5$.

The Rhind Papyrus

The Egyptians used algebra to solve practical problems. For instance, suppose one thousand stonecutters were building a pyramid, and each stonecutter ate three loaves of bread a day. How much bread would be needed to feed the stonecutters for ten days? The equation: $x = 1{,}000 \times 3 \times 10$.

Ancient Brain Teasers

Modern math textbooks sometimes include brain teasers and puzzles that are fun for students to solve. Brain teasers certainly are not a new idea. Teachers have been using puzzles to teach math for almost four thousand years. Here is one puzzle that Ahmes made up for Egyptian math pupils:

> Seven houses contain seven cats. Each cat kills seven mice. Each mouse had eaten seven ears of grain. Each ear of grain would have produced seven bushels of wheat. What is the total of all these [bushels of wheat]?

Strange Multiplying and Dividing

The Egyptians had a strange way of multiplying, which involved two columns of numbers. The left column always began with 1 and doubled with each row. The right column began with the number to be multiplied and doubled with each row. Suppose a student wanted to multiply 30 by 12. The student first would make two columns:

1	30
2	60
4	120
8	240
16	480

The student would then write down numbers from the first column that added up to 12: **4 + 8 = 12**. Then the student would add the "partners" of those numbers from the next column to get the answer: **120 + 240 = 360**.

Ancient Egyptians also divided with columns of numbers. Suppose a student wanted to divide 81 by 9. First, the student would make rows like those used in multiplication problems:

1	9
2	18
4	36
8	72
16	144
32	288

Then the student would find numbers in the right-hand column that added up to 81: **9 + 72 = 81**. Next, he or she would add the "partner" numbers to get the answer: **1 + 8 = 9**.

How Much Is Enough?

In ancient Egypt, scribes were important people. They kept records, figured taxes, managed building projects, and helped supply the military with food and equipment.

The Rhind and Golenishchev papyruses explained how to solve the kinds of problems that scribes would encounter in their work. The books had lessons on measuring the area of fields, adding up numbers of bricks, and calculating the amount of bread and beer needed to feed construction workers.

Measuring with Knots

The Egyptians used simple methods for measuring the size of fields and buildings. Sometimes they used wooden rods of uniform length to measure distances, much as we use yardsticks. Other times, they used knotted ropes. They tied knots in ropes, each knot placed an equal distance from the next. By running a rope along the ground or against the side of a building and then counting the knots, the Egyptians were able to determine length.

A drawing on a tomb built at Thebes around 1400 B.C. shows people using a knotted rope to measure a field of grain. In the picture, one man holds each end of the rope, which is stretched along the side of the field. Meanwhile, two other men record the measurement. The men look much like officials at modern football games using a 10-yard chain to measure the football's advance.

Great Surveyors

It took great surveyors to build the Great Pyramid at Giza, completed around 2560 B.C. The pyramid is 481 feet high and was built from more than two million stone blocks. Its base is 755 feet long on each side and covers an area bigger than 10 football fields. Yet the sides of the base come within seven inches of forming a perfect square. They are oriented almost exactly in north-south and east-west directions.

How did Egyptian surveyors work so accurately? Part of their secret was a tool called the *groma,* used to establish right angles (angles that are perpendicular—measuring 90 degrees). The groma was a flat wooden cross; its arms intersected one another to form four right angles. Cords were attached to each end of both arms, with weights tied to the ends of the cords. The weighted cords hung straight down, forming additional right angles with the arms of the cross. Ancient surveyors used the arms and cords as sights, lining them up with the walls and ceilings of buildings. The groma helped builders make sure that the walls formed perfect right angles with one another.

The Nilometer

The Nile's yearly flooding was important to Egyptian farmers. Too little flooding meant less water for crops and a bad harvest. Too much flooding could damage farms and cities.

Around 3000 B.C., the Egyptians created a device for computing the Nile's flood. Archaeologists call it the "Nilometer." The Nilometer consisted of stone gauges that recorded the depth of water when the Nile overflowed its banks.

Scribes and priests kept records of the floods for centuries. Year after year, they compared water levels to crop production. From these comparisons, scribes concluded that a flood of about 25 feet was best for the main Egyptian crops of wheat and barley.

Ancient Tax Reform

King Sesostris I, who ruled Egypt from 1962 to 1928 B.C., decided to charge his subjects a tax based on the amount of land that each person farmed. But every year when the Nile flooded, soil washed into the river, and some farmers lost big pieces of land. The farmers, however, had to pay the same amount of tax each year, even if they had lost land to flooding.

Pool and Nilometer at Dendera, Egypt

Sesostris was a fair king, and he ruled that farmers who had lost land could ask for a tax cut. Sesostris then sent scribes to measure the amount of land that had been lost. In searching for a fast way of figuring the loss, Egyptian scribes developed geometry, a form of math used to find the area of squares, rectangles, and other figures. The ancient Greek historian Herodotus described the process:

> Upon which, the king sent persons to examine, and determine by measurement the exact extent of the loss. Thenceforth only

> such a rent was demanded of him as was proportionate to the reduced size of his land. From this practice, I think, geometry first came to be known in Egypt, whence it passed into Greece.

The Grain: A Good Old Name

The smallest official unit of weight in the United States and the United Kingdom is the grain. A grain is tiny. It takes 437.5 grains to equal one ounce and 7,000 grains to equal one pound.

The ancient Egyptians first used the grain as a unit of measurement thousands of years ago. It originally equaled the weight of one grain of wheat. Merchants selling small amounts of precious goods, such as gold and incense, would

Three-thousand-year-old Egyptian sundial shadow clock

put several grains of wheat on one side of a scale and weigh out the goods on the other side.

Shadow Clocks, Sundials, and Water Clocks

People in ancient Egypt and other early civilizations cared about timekeeping. Priests and soldiers had to perform certain tasks at certain times. Rulers, government officials, and scribes had to keep track of workers and their time on the job. Like the Babylonians, the Egyptians divided both the day and the night into 12 equal parts.

The Egyptians used clocks as early as 3500 B.C. These were shadow clocks, consisting of a vertical stick that cast a shadow as the Sun moved past. The shadow grew shorter as the Sun rose in the east, disappeared at noon with the Sun directly overhead, and lengthened throughout the afternoon as the Sun dipped in the western sky.

Sometime around the eighth century B.C., the Egyptians made an improved shadow clock, a sundial like those used in the ancient Middle East, with lines representing times of day. The earliest known Egyptian sundial was made from a base of greenstone (a mineral also known as nephrite). Lines cut into the base stood for six units of time. A gnomon cast a shadow on one line after another as the Sun moved through the sky.

Around 1500 B.C., the Egyptians built a new timekeeper. It was the clepsydra, or water clock, made of a clay jar with markings on the inside. As water in the jar dribbled out of a small hole at the bottom, more and more markings were exposed. Each mark that appeared meant that another unit of time had passed. To make sure that clepsydras in different places all kept time the same, they had to be made very precisely—so that water flowed out of each one at about the same rate.

FOURTH OF JULY—IN DECEMBER?

The solar year is the time it takes Earth to make a full revolution around the Sun: approximately 365 days, 5 hours, 48 minutes, and 46 seconds. What would happen if the calendar didn't match the solar year? Holidays and whole seasons would gradually shift, with August eventually falling in the middle of winter.

The first ancient calendars did shift in this way. They were created according to the lunar year, which is divided into 12 months based on the phases of the Moon. The lunar year lasts only 354 days. Because of the difference between the lunar year and the solar year, the first calendars were not accurate. They shifted 110 days—almost four months—every 10 years.

The Egyptians were the first people to solve the problem. They created a calendar based on the solar year. The Egyptian calendar had 12 months of 30 days each, with 5 extra days added at the end of each year. In 238 B.C., King Ptolemy III made the calendar even more accurate by adding an extra day every fourth year. That day made up for the 5-hour, 48-minute, 46-second difference (about one-quarter of a day) between the calendar year and the solar year. The year with the extra day is called leap year.

chapter four

4

ANCIENT INDIA AND CHINA

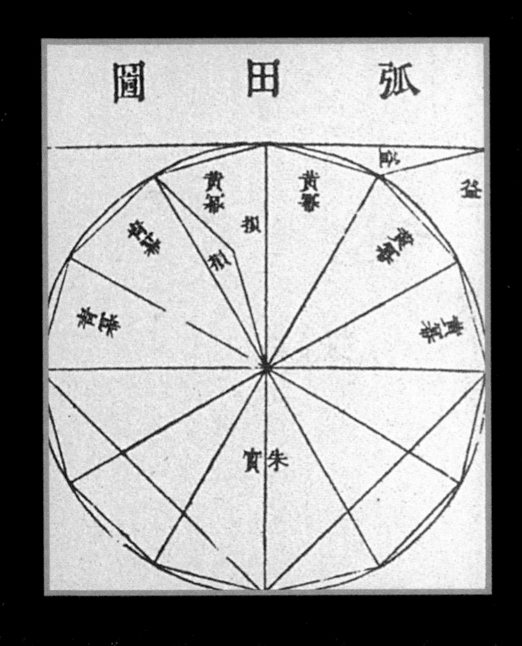

Ruins of the ancient Indus Valley city Mohenjo Daro. The mound in the background, called the Citadel, was the site of many important buildings.

People in the northwestern part of India began settling into villages around 4000 B.C. During the next thousand years, one of the world's greatest civilizations began to emerge there. We call it the Indus Valley civilization because it developed along the Indus River valley.

Historians were unaware of the Indus Valley civilization until 1922, when archaeologists found remains of ancient brick buildings and began to explore the area. Eventually, archaeologists uncovered remains of more than 50 cities and towns.

Evidence shows that the Indus Valley civilization lasted from 2500 to 1500 B.C., then disappeared. We still do not know why the civilization collapsed. But we do know that the ancient Indians introduced many advances in computing technology. For instance, the numerals used in the United States and most other countries developed in India. The ancient Indians also helped develop the decimal system.

Numerals	Mauryan Brāhmī BC 250 - 150	Post-Mauryan Brāhmī BC 150 - 0	Kṣatrapa Brāhmī AD 1 - 400	Gupta Brāhmī AD 400 - 550
1		—	—	—
2		=	=	=
3			≡	≡
4	+	[illegible]	[illegible]	[illegible]
5			[illegible]	[illegible]
6	[illegible]	[illegible]	[illegible]	[illegible]
7		[illegible]	[illegible]	[illegible]
8			[illegible]	[illegible]
9		[illegible]	[illegible]	[illegible]
10		[illegible]	[illegible]	[illegible]
20		O	θ	θ ⊙
30			[illegible]	[illegible]
40			[illegible]	
50	[illegible]		[illegible]	
60			[illegible]	[illegible]
70			[illegible]	[illegible]
80		[illegible]	[illegible]	[illegible]

Left: Indian numerals in different forms, as known from ancient coins dated between 25 B.C. and A.D. 550, next to our modern numeral equivalents. Below: Indian numerals from A.D. 600

Arabic Numerals = Indian Numerals

The credit for major technological advances sometimes gets lost in history. Sometimes great discoveries are named after the wrong person—or the wrong civilization. That certainly is the case with the numerals used in most of the modern world. We call them Arabic numerals, but ancient Indian people actually developed them.

The Indian numeral system allowed people to write down any number, no matter how big, with just 10 symbols: 0, 1, 2, 3, 4, 5, 6, 7, 8, and 9. The first known images of these numerals appear on pillars built by Asoka, an Indian king, around 250 B.C.

Middle Easterners learned about the numerals by trading with India, and they adapted the system for their own use. Around A.D. 976, Europeans learned the system from the Middle East. Europeans didn't know about the numerals' roots in ancient India, so they named them Arabic numerals (from *Arab,* a generic name for Middle Easterners), a term that still stands.

The Power of Zero

An unknown Indian mathematician made the first known reference to the concept of zero in 876 B.C. He suggested that an unused row on an abacus, an ancient counting device, be designated by a special symbol. Many centuries passed before zero went into wide use, however.

When zero did come into common use, it was an important advance in computing. Zero is necessary to the place-value system, which enables numbers to hold different values, depending on their placement. Think about the number 222. Each 2 has a different value, depending on its position in the number. The first 2 stands for 200. The second stands for 20. The third stands for 2. Zero and the place-value system enable us to write very large numbers (74,001) and very small ones (0.0471) with only 10 symbols—0 through 9.

The place-value system made adding, subtracting, multiplying, and dividing easy. Numbers could be written one under the other in columns and lined up according to value. Adding and subtracting became a snap. (Just try adding or subtracting large numbers with the numerals used in ancient Egypt. It would make you scream.)

Math and Religion

The Hindu religion, first practiced in ancient India, played a big role in encouraging advanced mathematics in Indian society. The sacred books of Hinduism said that every male head of a family had to perform special acts of worship, called *purvas,* each day.

Altars for worship had to be built in precise shapes and sizes described in the sacred books of Hinduism. For instance, one altar might have to be circular and measure a certain area. The next altar might have to be a square equal in area to the circular altar. A third might have to be exactly three times the area of the first altar. People in ancient India needed math in order to pray! But since most ordinary Indian people did not know how to calculate the area of geometric figures, scholars and priests usually performed the mathematics.

Ancient Indian scholars made calculations on small computing boards. Some boards were square, about 12 inches on each side, and covered with a thin dusting of red powder.

Numbers were scratched into the powder with a thin, pointed stick. Other boards were more like small chalkboards. Numbers were written with a pen dipped in white ink that could be easily rubbed off. Making corrections and erasing calculations was simple.

FATHER OF SINE

The name *trigonometry* comes from Greek words that mean "the measurement of triangles." At the very heart of this branch of mathematics are six functions, or ratios, used to determine the sizes of the sides and angles of triangles. One of these six trigonometric functions is called "sine."

An ancient Hindu mathematician, Aryabhata the Elder, computed the first sine tables, lists showing the value of sine for angles of many sizes. The tables let mathematicians perform trigonometry fast, without stopping to figure out the sine for each angle. Aryabhata included the tables in a book he published in 499 B.C. In addition to the sine tables, the book included many other advances in computing and mathematics.

ANCIENT CHINA

The ancient Chinese were good mathematicians. They developed the basis for a decimal system and invented the first automated computing device, the abacus.

Won-wang (1182–1135 B.C.) is the earliest Chinese mathematician known to us by name. He wrote the first Chinese essay on math. It was called the *I-king,* or *Book of Permutations.*

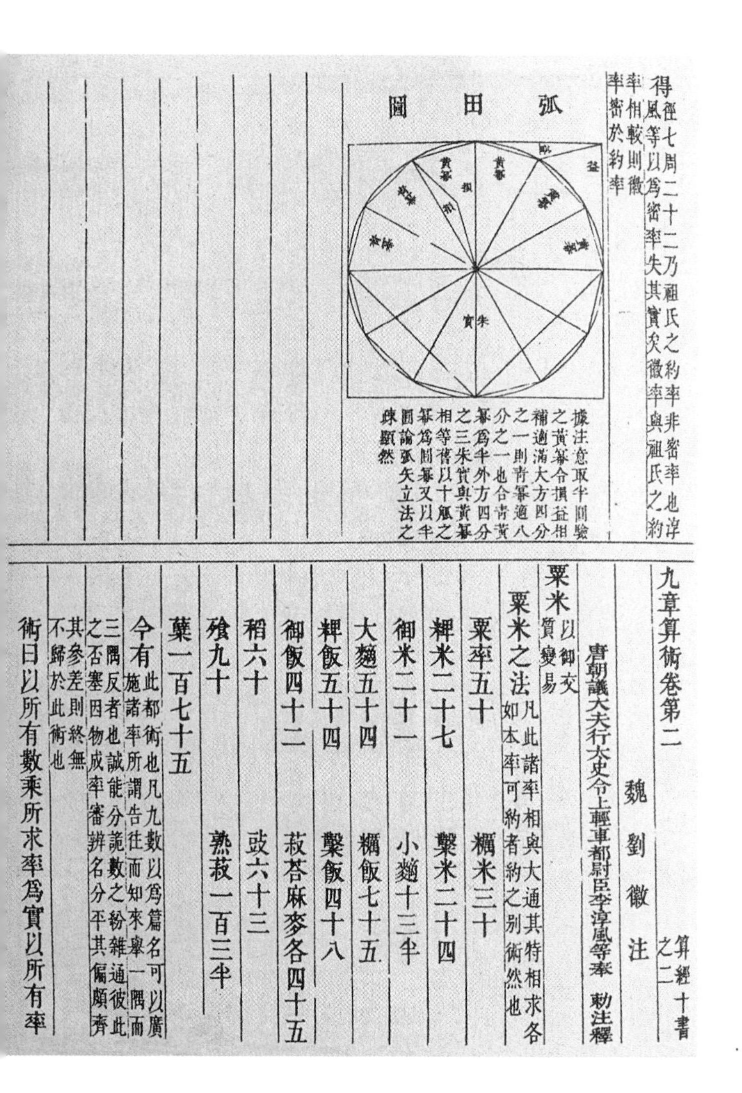

得徑七周二十二乃祖氏之約率非密率也淳
風等以爲密率失其實矣徽率與祖氏之約
率相較則徽率密於約率

弧田圖

據注意取半圓驗之黃冪合損益相補適滿大方四分之一則青冪適八分之一也合青黃冪爲半外方四分之三朱實與黃冪相等舊以十觚之冪爲圓冪又以半圓論弧矢立法之疎顯然

九章算術卷第二 算經十書之二

魏 劉 徽 注

唐朝議大夫行大史令上輕車都尉臣李淳風等奉 勅注釋

粟米 以御交質變易

粟米之法 凡此諸率相與大通其特相求各如本率可約者約之別術然也

粟率五十 糲米三十

粺米二十七 糳米二十四

御米二十一 小䵂十三半

大䵂五十四 糲飯七十五

粺飯五十四 糳飯四十八

御飯四十二 菽荅麻麥各四十五

稻六十 豉六十三

飧九十 熟菽一百三半

糵一百七十五

今有 此都術也凡九數以爲篇名可以廣施諸率所謂告往而知來舉一隅而三隅反者也誠能分詭數之紛雜通彼此之否塞因物成率審辨名分平其偏頗齊其參差則終無不歸於此術也

術曰以所有數乘所求率爲實以所有率

Reproduction from *Nine Math Chapters,* an ancient Chinese math document. The document was compiled by two mathematicians during the West Han dynasty (206 B.C. to A.D. 24).

STICK NUMBERS

The ancient Chinese wrote several different kinds of numerals, each for a different purpose. "Official" numerals were used on contracts and other business documents. Stick numerals, consisting of short lines, were used for mathematics. In stick numerals, zero was written as a small square. The numbers 1 to 5 were represented with the same number of sticks. T-shaped figures were used to show the numbers 6 to 9.

Math students and scholars in ancient China performed calculations on counting boards using sets of red and black counting sticks. The boards were made from wood and were separated into squares, sort of like large modern chessboards. Each counting stick was about four inches long and looked like an old-fashioned matchstick. A full set contained 271 sticks.

Sticks were placed on the counting board's squares from left to right, with each square representing the 1s, 10s, 100s, 1,000s, and so forth. An empty square stood for zero. Red sticks were used for numbers to be added and black sticks for numbers to be subtracted.

Chinese scholars used counting sticks with amazing speed. They threw the sticks down, glanced at and memorized the answers to equations, swept the boards clean, and started the next problem in the wink of an eye.

Ancient Computers

Ancient Chinese civilization reached its peak with the Chou dynasty (the period during which the Chou family ruled), which lasted from about 1122 to 256 B.C. During this period, the Chinese developed the world's first automated computing machine—the abacus. It was in widespread use in China by 500 B.C.

The abacus could be considered the world's first computer. It was used for addition, subtraction, multiplication, and division. With the abacus, people could perform these

Nineteenth-century A.D. Chinese abacus

calculations much faster than they could with counting boards or numbers written on paper.

The abacus consisted of a rectangular frame with a series of vertical rods. Beads slid up and down along the rods. A crossbar divided the frame into two decks. The upper deck had two beads on each rod. The lower deck had five beads on each rod.

Computing with an abacus was simple. With the device flat on a table, the user moved and counted the beads. Each bead in the lower deck had a value of 1. Each bead in the upper deck had a value of 5. A bead was "counted" when it was moved to the crossbar separating the decks.

Electric adding machine from the 1940s

Ancient Versus Modern Technology

In 1946 in Japan, the U.S. Army newspaper, *Stars and Stripes,* sponsored a contest between ancient and modern technology. The newspaper found the army's most expert operator of an electric adding machine. At that time, the adding machine was the most advanced computing device in everyday use. *Stars and Stripes* also found the most expert Japanese abacus operator.

Each man was given the same problems in addition, subtraction, multiplication, and division. These problems included adding 50 numbers, each containing three to six digits, and dividing numbers with 5 to 12 digits in the divisor and dividend. Referees used stopwatches to see which technology was faster. Here is how the Japanese newspaper *Nippon Times* reported the outcome:

Civilization, on the threshold of the atomic age, tottered Monday afternoon as the two-thousand-year-old abacus beat the electric calculating machine in adding, subtracting, dividing, and a problem including all three with multiplication thrown in. . . . Only in multiplication alone did the [electric] machine triumph.

Each vertical rod represented a place value—the 1s, 10s, 100s, 1,000s, and so on. To show the number 4,321 on an abacus, the user moved one bead on the far right rod of the lower deck to the crossbar, two beads on the next rod, three on the next, and four on the next.

When five beads on a rod had been counted, the user "carried" the number to the upper deck, moving one of the upper beads to the crossbar and all five lower beads away from it. When both upper beads on a rod had been counted, the user carried the number to the next rod by moving one of the lower beads toward the crossbar.

Best for Thousands of Years

After the invention of the abacus, no major advances occurred in automating mathematics or counting for two thousand years. The next major breakthrough came in 1642, when French mathematician Blaise Pascal built the first mechanical adding machine.

Even at the start of the twenty-first century, merchants and store clerks in China, Japan, and other countries still use abacuses. Some abacus operators can perform calculations faster than a person with a handheld calculator. What's the secret? The operator's skill in inputting data. Skilled abacus operators can move beads faster than most people can hit the keys on a calculator.

chapter five

5

ANCIENT AMERICAS

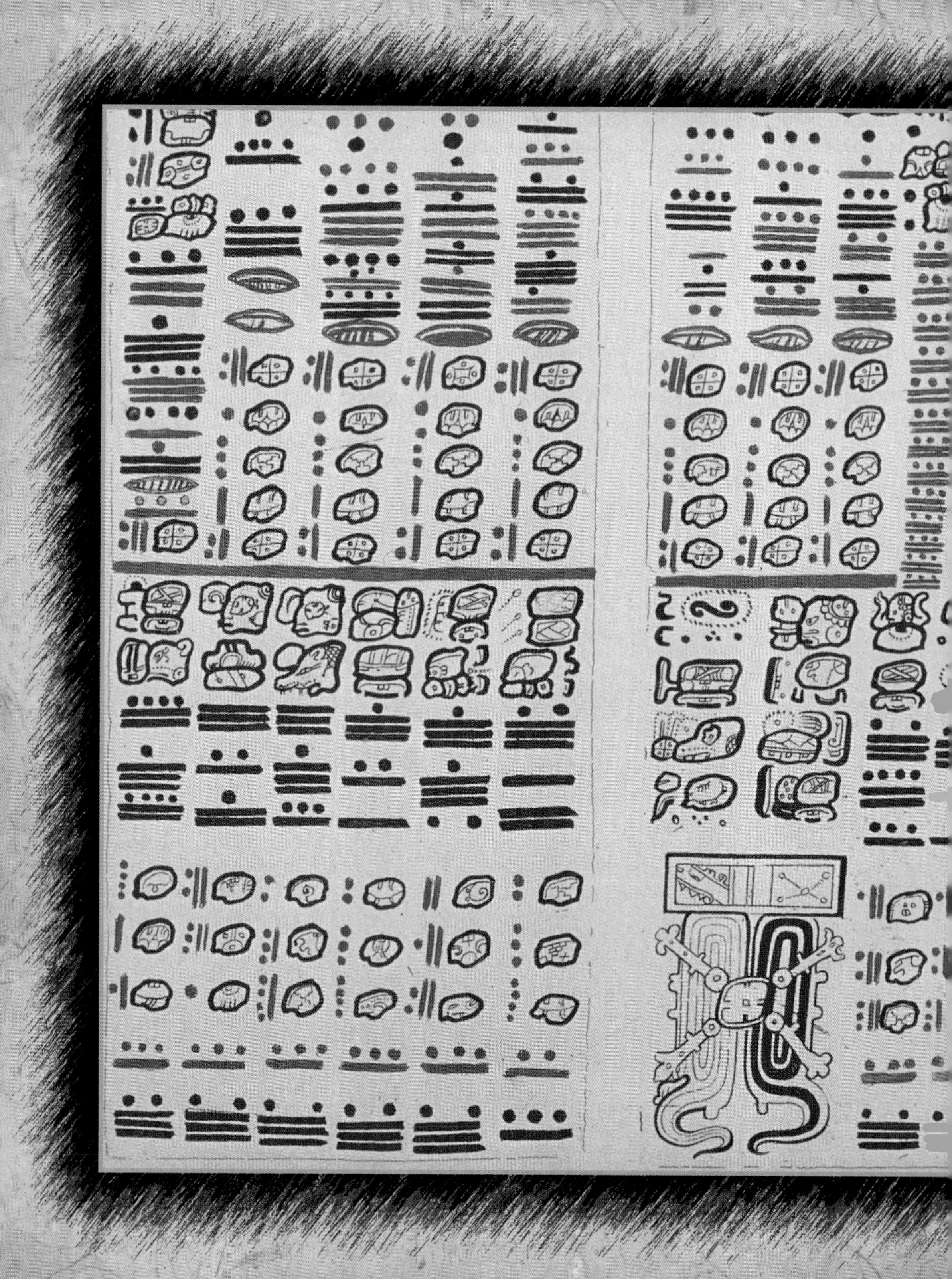

in powers of 20: 1s on the bottom row, 20s above them, 400s next, and so on.

Holy Numbers!

The Maya believed that some numbers were holy, including 20, the basis for the Maya numbering system. The number 5 was also special, maybe because people have five fingers on each hand. The Maya century had 52 years, so the number 52 was holy. The number 400 was very special—it was the number of gods of the night. The number 13 was special because the Maya thought that the Earth and sky consisted of 13 layers. Maya priests counted 1,600 stars in the sky, so 1,600 was also a sacred number.

The Maya Calendar

The Maya believed that some days were lucky—and some unlucky. Lucky days were chosen for important activities such as weddings, battles, and planting and harvesting crops. The Maya also celebrated many religious holidays, which had to be observed on the same days each year. To keep track of lucky days and holidays, the Maya needed an accurate calendar—one in step with the solar year.

The Maya used two calendars. The Tzolkin, or sacred calendar, contained 260 days. The days were named after Maya gods, thought to carry time across the sky. The Tzolkin was used in combination with the Haab, a 365-day calendar based on the solar year.

The Maya calendars were very complicated. They were based on movements of the Sun, the Moon, and the planet Venus. Scientists do not know exactly how the Maya calendars were developed, but they came into use in the first on. The Maya built the number 62 with three 20s and two 1s. They built the number 184 with nine 20s and four 1s. We don't know why the Maya chose the base-20 system—perhaps because people have 20 fingers and toes in all.

Maya numerals consisted of dots and bars. One dot stood for 1, two dots stood for 2. One bar stood for 5, two bars stood for 10. The Maya combined dots and bars to write larger numbers. One bar and four dots, for instance, stood for 9. Two bars and two dots stood for 12.

Nothing Counts—A Lot

Zero. 0. Nothing. Naught. It's hard to imagine modern mathematics without the symbol 0. With zero and 1 though 9, people can express any other number. With just a few digits, 0, 3, 4, and 5, for instance, we can write 3, 30, 304, 340, 403, 503, 4,000, 4,003, and thousands of other numbers.

The ancient Maya began using a symbol for zero in the second century A.D.—long before Old World civilizations used such a symbol. The Maya symbol for zero was a small oval with several lines inside. The first known zero in the Old World was found on a copper plate from Gujarat, India, and dates to around A.D. 585.

Numbers All in a Row

The Maya numbering system depended on position, just like our modern place-value system. Our numbers increase in value from right to left, with 1s in the right-hand position, 10s in the next spot, then 100s, and so on.

Maya numerals were written from bottom to top. And because the Mayans used a base-20 system, values increased base-10 system is made with six 10s and two 1s. The number 184 consists of one 100, eight 10s, and four 1s.

The Maya used a system based on the number 20. It was a vigesimal system, named from the Latin word *vigesimus,* meaning "twentieth." In the Maya system, place values increased in powers of 20: 1, 20, 400, 8,000, 160,000, and so

Maya hieroglyphic writing, including numbers, from A.D. 600. The Maya sometimes used head-shaped symbols, rather than bar-and-dot numerals, to represent numbers. The number one, for example, was often depicted as the head of a young earth goddess. This is similar to using the word *one* instead of the numeral *1*.

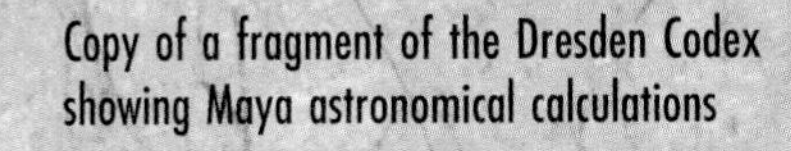
Copy of a fragment of the Dresden Codex showing Maya astronomical calculations

The Maya, a New World civilization based in Mesoamerica (Mexico and Central America), began settling into farming villages in 1200 B.C. or earlier. Eventually, Maya towns and cities stretched from southern Mexico through the modern-day countries of Honduras, Guatemala, and Belize.

Maya computing technology included a numeral system, mathematics, and an accurate calendar. Maya computing was more advanced, in some respects, than math used at the same time in the Old World. Specifically, Maya mathematics included the concept of zero, which was not widely used in the Old World until after ancient times.

Counting by 20s

Most modern societies use a system of counting based on the number 10. With this base-10 system, place values increase in powers of 10: 1, 10, 100, 1,000, 10,000, 100,000, 1,000,000, and so on. For example, the number 62 in the

chapter six

6

ANCIENT GREECE

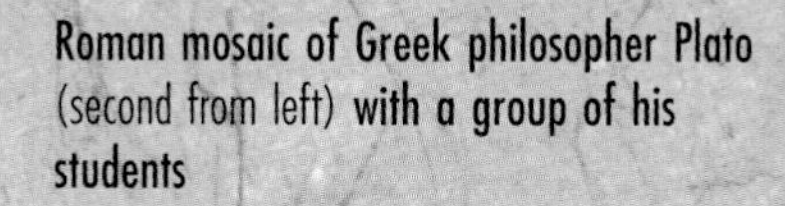
Roman mosaic of Greek philosopher Plato (second from left) with a group of his students

Ancient Greece was a powerful civilization that conquered much of the Mediterranean and Middle Eastern world—from Egypt to India. In Egypt, the Greeks founded the city of Alexandria, which became a center for computing and science. The Greeks borrowed some computing technology from the Egyptians, but they did not just make small improvements. Instead, the Greeks developed entirely new fields of computing—laying the foundation for modern mathematics.

To the Egyptians, math was a practical tool used for figuring taxes, conducting business, and building structures. The Greeks, on the other hand, admired math for its logic. They thought of it as a way to train the mind.

The Greeks separated math into two main branches: applied math, used to solve practical problems, and theoretical math, the study of lines, figures, and points that do not exist in nature. The Greeks also used math to prove and disprove theories about the natural world.

Pythagoras

The Magic Square

People in ancient Rome, China, India, and some other civilizations used numbers as good-luck charms. The followers of Pythagoras, a famous Greek mathematician, thought that numbers were the source of all order and harmony in the universe.

Some groups of numbers were called "magic squares." They were numbers arranged in a square, so that the sums of all columns, rows, and diagonals were equal. In the following magic square, each column, row, and diagonal totals 15:

2	7	6
9	5	1
4	3	8

A drawing of one magic square was found in the ruins of Pompeii, a Roman city that was destroyed in A.D. 79 by the eruption of Mount Vesuvius.

Greek Numerals

Imagine having to memorize 27 symbols for numbers instead of the 10 we use. That's what students in ancient Greece did. The Greeks used the 24 letters in their alphabet to stand for numbers. When they ran out of their own letters, they borrowed three letters from the Phoenician alphabet.

The first nine Greek letters stood for the single-digit numbers, 1 through 9. The second nine letters represented 10, 20, 30, and so on up to 90. The last nine letters stood for 100, 200, 300, and so on up to 900. A bar placed to the left of a numeral indicated thousands. The letter *M* below a numeral stood for tens of thousands.

A Famous Theorem

Pythagoras of Sámos, who lived from about 580 to 500 B.C., lived in Crotona in southern Italy, where he started a school of math and philosophy. His students were called the Pythagoreans.

Pythagoras is best remembered for devising what may be the most famous theorem (a statement that has been proved or is to be proved) in mathematics: "The square of the length of the hypotenuse of a right triangle is equal to the sum of the squares of its legs." We often state the Pythagorean theorem as $a^2 + b^2 = c^2$. In this equation, a and b stand for the legs of the triangle, and c stands for the hypotenuse.

Although Pythagoras gets the credit for the theorem, the Babylonians knew about this equation a thousand years before Pythagoras. The Babylonians used the equation in surveying land and figuring the area of fields.

One Theorem Leads to Another

Euclid, another Greek mathematician, taught math in Alexandria. He studied prime numbers (a prime number is one that can be divided evenly only by 1 and itself) and proved that there are an infinite number of primes.

Around 300 B.C., Euclid organized all the major mathematical theorems, using one theorem to prove another and that theorem to prove the next. But Euclid ran into a problem. If each theorem was proved with an existing theorem, how could a person prove the *first* theorem? Euclid solved that problem using axioms—statements so obvious that proving them is unnecessary. Here are a few axioms that Euclid used:

- Things that are equal to the same thing are equal to one another. Example: **3 + 1 = 4** and **2 + 2 = 4**; therefore **3 + 1 = 2 + 2**.
- If equals are added to equals, the results are equal. Example: **(3 + 1) + 4 = (2 + 2) + 4**.
- If equals are subtracted from equals, the remainders are equal. Example: **(6 – 1) – 4 = (7 – 2) – 4**.
- Things that coincide with one another are equal to one another. Example: **6 + 4 = 6 + 4**.
- The whole is greater than the part.

With axioms and theorems, Euclid organized a system of geometry that is called Euclidean geometry in his honor. Euclid put his system into a 13-volume book, *The Elements,* which was used as a basic geometry textbook for two thousand years. Modern high school geometry courses are still based on the first volumes of *The Elements.*

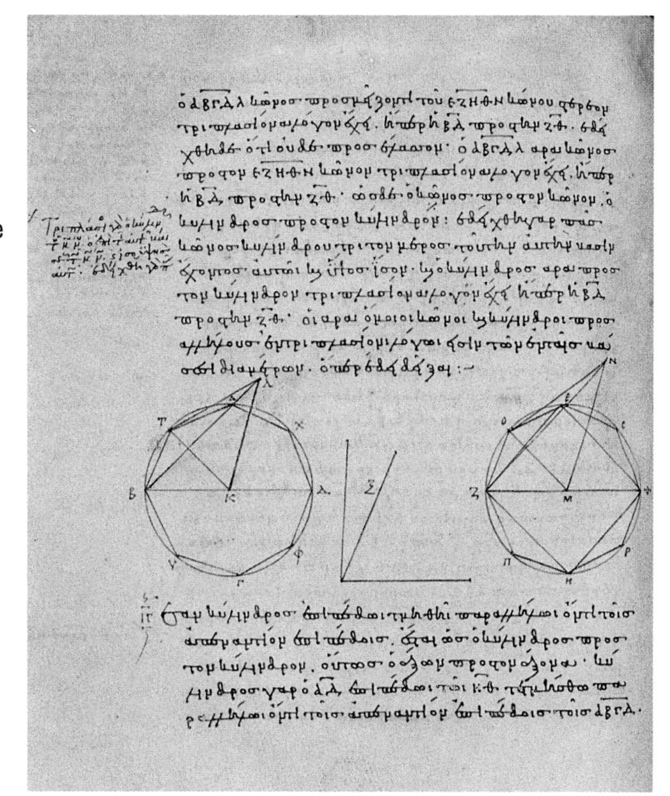

Copy of a page from volume 12 of Euclid's *The Elements,* showing his geometric diagrams. This copy, called the *D'Orville Euclid,* was drafted in A.D. 888.

EVEN A DUMB ANIMAL WOULD KNOW

Although Euclid was one of the greatest mathematicians in history, experts think some parts of *The Elements* are silly. They say Euclid wasted lots of time proving ideas that were obvious and needed no proof.

In one part of *The Elements,* for instance, Euclid proved that no one side of a triangle can be longer than the other two sides added together. He drew a triangle with the corners labeled A, B, and C. If a hungry ass stood at point A, and a bale of hay sat at point B, Euclid explained, the ass would know that the shortest route to the hay was directly from point A to point B, not from A to C to B. Other mathematicians made fun of Euclid for spending time proving something that was obvious even to an animal.

Ancient Computing School

The Greek philosopher Plato lived from about 427 to 347 B.C. He was a great thinker who believed that society would benefit if everybody were educated to the highest level possible.

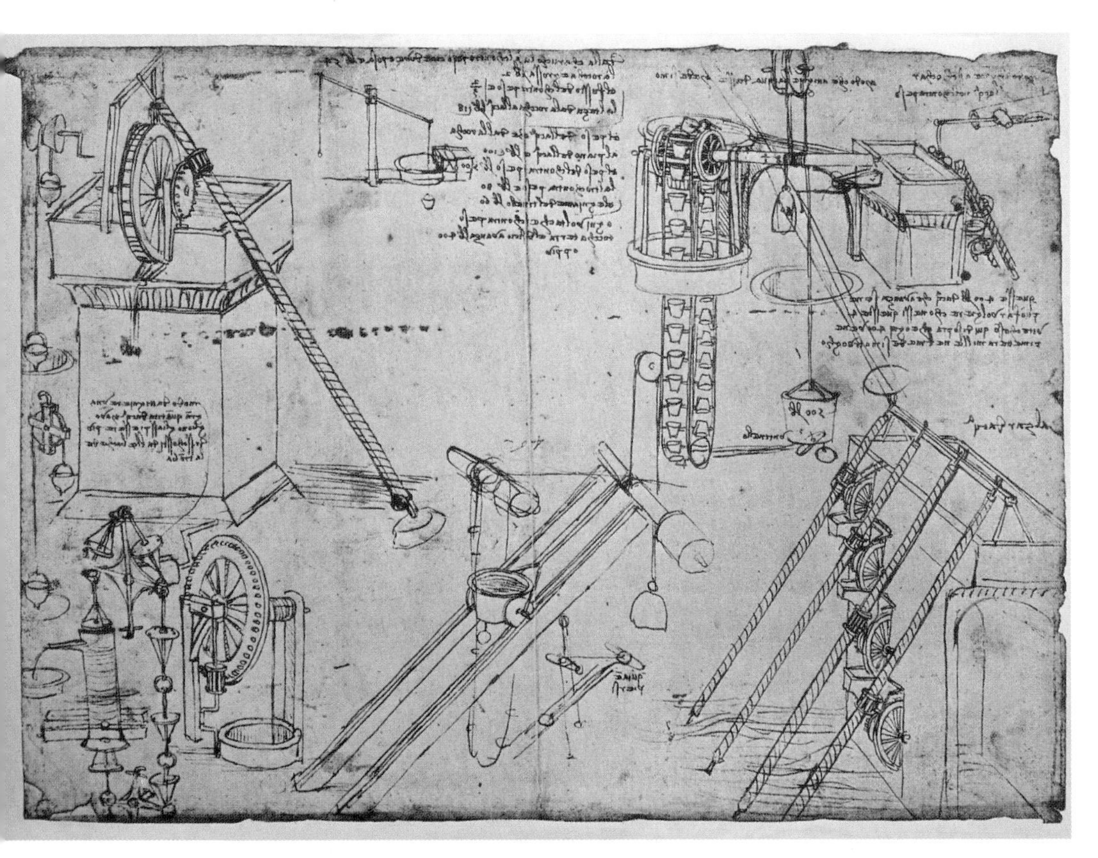

Italian engineer Leonardo da Vinci's sketches of some of Archimedes' designs. Da Vinci created these sketches between A.D. 1503 and 1507.

Although Plato was not a mathematician, he loved math, and he used his fame to encourage people to study it. Many great mathematicians came to Plato's Academy, a school of philosophy and science, in Athens. Above the school's doorway, Plato put a sign: "Let no one ignorant of mathematics enter here."

Eureka!

Archimedes, who lived from about 290 to 211 B.C., was the science adviser to Hiero II, the ruler of Syracuse in Sicily. Hiero asked Archimedes to find out whether his crown was made of pure gold or a mixture of gold and silver. Archimedes figured out the answer while sitting in the bathtub. He leaped from the tub and ran naked through the streets shouting *"Eureka! Eureka!"*—Greek for "I have found it."

Archimedes had discovered that an object submerged in water would displace, or push aside, water equal to its own weight. For instance, a ship will sink into the ocean just far enough to replace an amount of water equal to the ship's weight. Since gold weighs more than silver, Archimedes knew a crown of pure gold would displace more water than a similar crown of gold and silver. He used his discovery to figure out that the king's crown was not made of pure gold.

The Greatest Mathematician?

Some historians think that Archimedes was the ancient world's greatest math wizard. He used math to design machines and made great advances in computing technology. For example, Archimedes computed a new value for pi that was much more accurate than earlier figures. Mathematicians used his figure for centuries.

Archimedes studied spirals and figured out properties such as the surface area covered by each of a spiral's turns. In his study of spirals, Archimedes developed special math techniques. They were the basis, centuries later, for a field of math called calculus.

Archimedes also studied the properties of spheres and cylinders. He wrote about his discoveries in books such as *On Spirals*, *On Floating Bodies*, *The Measurement of the Circle*, and *On the Sphere and Cylinder.*

In one of his last works, *The Sand-Reckoner,* Archimedes calculated how many grains of sand it would take to fill the universe. He ended up with a number equal to 10 followed by 63 zeros. To find this figure, Archimedes developed a system for computing with very large numbers. It was the basis for scientific notation, a way of writing very big—and very small—numbers. In scientific notation, 894,000,000 would be written 8.94×10^8, where 10^8 is the same as a 1 followed by eight zeros.

Father of Algebra

Diophantus was a mathematician who lived in Alexandria around A.D. 275. Diophantus has been called the Father of Algebra. He advocated the use of symbols and equations (such as $x + y = z$) in math and wrote *Arithmetica,* the world's first book on algebra.

Arithmetica consisted of 13 volumes and included about 130 problems. It was used for centuries, and it helped engineers use algebra, not just geometry, in measuring land and constructing roads and buildings.

A Space Monster Is Eating the Sun: Part Two

Remember how Mesopotamian astronomers used math to study the movements of heavenly bodies and to predict eclipses? In the second century B.C., a Greek astronomer named Hipparchus of Rhodes also studied astronomical events.

Hipparchus learned about the Mesopotamian system of counting seconds and minutes by 60s and used it to track the positions of heavenly bodies. He also used trigonometry to study the position of the Sun and planets and was able to more accurately measure the length of a year. His calculation was correct to within 6.5 minutes.

A Famous Woman's Life

The first female mathematician known by name was Hypatia. Her father, Theon of Alexandria, was also a great mathematician.

Born around A.D. 370, Hypatia probably learned math and geometry from her dad. She went on to become the leader of a large group of philosophers and scientists based in Alexandria, Egypt. Hypatia wrote a number of works on mathematics, including a book on astronomy.

An angry mob murdered Hypatia in A.D. 415. They tore her body apart and scattered it throughout Alexandria. Why? Nobody knows exactly. Some historians believe that Hypatia became too famous and that people envied her. Others think that the murder was part of a backlash against scientists. Some people in ancient times thought that scientists did not believe in God. The mob may have killed Hypatia for that reason.

Many mathematicians and scientists fled Alexandria after

Statue of Hypatia made around A.D. 400

Hypatia's murder. They probably feared that they, too, would be killed. Before the murder, Alexandria had been the world's center of science, medicine, and learning for almost seven hundred years. Historians use Hypatia's death to mark the end of Alexandria as the center of learning.

Tick Tock . . . Better Clocks

A great Greek engineer named Ktesibios of Alexandria made an early form of our modern mechanical clock in the second century B.C. It was a clepsydra, or water clock, more advanced than those used in ancient Egypt. Ktesibios's clock consisted of a float attached to a vertical rod. A doll-like statue, with a pointer in its hand, was attached to the rod's top. The float was placed in a container of water. As water flowed out a hole in the bottom of the container, the float moved down. The pointer also moved down, marking the passage of time.

Other Greek water clocks were fancier. Their pointers were attached to gears that moved as water flowed from the clocks. The gears' movements caused polished stones to drop into bowls or little statues to spin, marking the passage of time.

Ancient Clock Tower . . .

Have you ever heard of Big Ben? It's a 13-ton bell in the Clock Tower of the Houses of Parliament in London, England. People have used Big Ben to keep track of the time since 1859. The Tower of the Winds in Athens held giant sundials that were the ancient world's Big Ben. Also known as the Horologium, the Tower of the Winds is still standing. It is a marble tower 42 feet high and 26 feet in diameter.

Andronicus of Cyrrhus, an astronomer and mathematician, designed the tower around 100 B.C. It has eight sides, each containing a sundial that keeps very precise time. Using geometry, Andronicus computed exactly how shadows would fall onto the surfaces of the sundials. For telling time at night and on cloudy days, Andronicus added a clepsydra to the tower.

. . . Complete with Ancient Computer?

In 1901 divers swimming off Antikythera Island, near Greece, found the remains of a strange mechanical device.

The Tower of the Winds in Athens, Greece

It came from the wreck of a ship that had sunk two thousand years earlier. The device became known as the Antikythera mechanism.

Nobody knew what the device was until the 1950s, when a Yale University scientist, Derek de Solla Price, concluded that it was an ancient computer. It had gears, pointers, and dials that turned to calculate the rising and setting of the Sun and Moon and movements of important stars. Professor de Solla Price thought that the Antikythera mechanism had probably been displayed in a museum or public hall, where people could look at it and marvel. Maybe, he said, it had been displayed in the Tower of the Winds. No one developed a more advanced computing device than the Antikythera mechanism for more than a thousand years after its creation.

Hero's Gift to Computing

In 1642 a French mathematician named Blaise Pascal invented the first mechanical adding machine. Pascal built it to help his father, a tax collector who had to add lots of figures. Pascal's machine could add and subtract faster than any human making calculations by hand.

Pascal got the idea for his adding machine from ancient technology. He had read about an odometer, a device for measuring how far a vehicle traveled, designed in A.D. 2 by Hero of Alexandria. Hero's odometer consisted of a gear connected to the axle of a chariot. Each turn of the axle turned the gear, which triggered a counter. The counter kept track of how far the chariot traveled.

Pascal used the same basic mechanism in his machine to add numbers. The same kind of machine was used throughout the twentieth century in odometers and meters that recorded the use of water, natural gas, and other materials.

The Motionless Runner

Computing technology can do more than solve practical problems. It can help prove or disprove ideas about the natural world. The Greek philosopher Zeno of Elea, who lived around 450 B.C., used math to try to prove other philosophers' theories about time and space.

His ideas became known as Zeno's Paradoxes. A paradox is a statement that is contradictory or illogical. The "Paradox of the Motionless Runner" was Zeno's most famous:

> Motion cannot exist because before that which is in motion can reach its destination, it must reach the midpoint of its course, but before it can reach the middle, it must reach the quarter point, but before it reaches the quarter point, it must first reach the eighthpoint, etc. Hence, motion can never start.

To understand the paradox, suppose a girl wants to run to a friend 100 feet down the street. First she has to reach the 50-foot mark. Before that, she has to reach the 25-foot mark. To do that, she must reach the 12.5-foot mark. Before that, she has to run 6.25 feet. Since space can be divided into smaller and smaller units infinitely—without end—the girl can never reach her goal.

This paradox may seem silly or illogical. But it created a big headache for mathematicians. For centuries, no one could prove Zeno wrong. Mathematicians finally succeeded in the mid-1850s. They disproved the paradox using the theory of infinite sets, a concept you may study in high school or college.

chapter seven

7

ANCIENT ROME

Roman mosaic of dice players

Ancient Rome began as a small town, founded in 753 B.C. and located on the Tiber River in central Italy. The town's early settlers were called the Latins, one of many groups then living in Italy. Rome was first ruled by the Etruscans, who controlled the surrounding area. They built Rome into a great city. In 509 B.C., the Roman people overthrew the Etruscans and established a republic, a government ruled by the citizens rather than a king.

Gradually, the Romans conquered neighboring lands and built a great empire. It eventually extended from the Caspian and Red Seas in the east to Spain in the west and to England in the north. Like many other ancient civilizations, the Romans learned about technology from neighboring groups, including the people they conquered.

In fact, the Romans inherited much of their computing technology from the Greeks. But the Romans used math and computing in a much different way. Whereas the Greeks admired math as a way of training the mind, the Romans were practical people who applied math to real-life problems. They needed freshwater for their cities, so they used computing technology to design aqueducts. Using the groma and other surveying instruments borrowed from the conquered Egyptians, the Romans constructed roads, buildings, and other structures. The Romans used math as a tool—but they made few important mathematical advances of their own.

I, V, X, L, C, D, M

The Romans developed a numbering system in which just seven letters from the Latin alphabet could be used to write any number—from 1 to 1,000,000,000,000,000,000,000,000 or more! The Latin letter *I* stood for 1, *V* for 5, *X* for 10, *L* for 50, C for 100, *D* for 500, and *M* for 1,000.

Placing a small bar over the top of a number multiplied its value by 1,000. For instance, M (1,000) with a bar on top meant 1,000,000. In theory, a person could add enough bars to write huge numbers. In practice, the Romans rarely used more than one bar.

Roman numerals were written from left to right. A number placed to the right of another number of equal or greater value indicated addition. That is, VI meant 5 + 1, or 6. MD meant 1,000 + 500, or 1,500. DC meant 500 + 100, or 600. A number placed to the left of another number of greater value indicated subtraction. For example, XL meant 50 – 10, or 40. DM meant 1,000 – 500, or 500. MCM meant 1,000 + 1,000 – 100, or 1,900.

Try writing your street number, height, and weight using Roman numerals. Can you guess their big disadvantage?

For one thing, the numbers can take up a lot of space. They are also hard to use for addition, subtraction, multiplication, and division.

How Far Have We Gone?

Remember Hero of Alexandria's odometer, designed in A.D. 2 to measure how far a vehicle had traveled? The first odometer was actually developed a number of years earlier, by Marcus Vitruvius Pollio, a Roman engineer who lived from 70 to 25 B.C.

He mounted a large wheel in a frame, much like the base of a wheelbarrow. The wheel was attached to a gear with four hundred notches, and with each turn of the wheel the gear moved ahead one notch. The gear moved four hundred times every five thousand feet—equal to one Roman mile. And with each four hundred turns, a stone dropped into a metal container. The clang of the stone signaled that one mile had passed.

Vitruvius envisioned using the device in chariots and wagons so that travelers could measure distance. At the end of each day's travel, a driver could count the number of stones in the container, tally his mileage, and reload the stones for the next day's journey. The Romans never adopted the odometer, however. It was forgotten for more than 1,300 years, until the great Italian engineer Leonardo da Vinci (1452-1519) discovered Vitruvius's written description.

Better Calendars

The first Roman calendar, developed around 738 B.C., was based on the lunar year. With just 304 days—divided into

Julian calendar with months, days of the week, and the Roman zodiac. At the right and left sides are Roman numerals.

10 months—the calendar was 61 days short. Later, the Romans added two more months, but they were not long enough. To make up for the shortfall, the Romans had to add an extra month to their calendar every two years.

The Roman calendar got further off track when officials started adding even more extra months. Why? Sometimes they did it so they could stay in office longer or delay elections. Finally, in 45 B.C., Emperor Julius Caesar adopted the Egyptian solar calendar, calling it the Julian calendar. It had a year of 365 days and every fourth year—leap year—an extra day. The Julian calendar was very accurate—only 11 minutes, 14 seconds longer than the solar year.

People in many parts of the world used the Julian calendar for more than 1,500 years. But by 1582, the 11-minute error had accumulated to make 10 days. Holidays were occurring at the wrong time of year. So Pope Gregory XIII ordered that 10 days be dropped from the calendar. He adopted a new calendar, known as the Gregorian calendar, which was almost exactly as long as the solar year.

The Gregorian calendar used the designations B.C. and A.D., based on the birth of Jesus Christ. B.C. stands for "Before Christ." A.D. stands for the Latin words *anno Domini*, "In the year of our Lord."

The End of Ancient Computing

Advances in computing ended after the Western Roman Empire collapsed in A.D. 476. Knowledge of algebra, geometry, and other branches of math faded.

Boethius, a Roman philosopher who wrote simple mathematics textbooks, died in A.D. 524. Historians say that his death ended mathematics in Europe. "From the years 500 to 1400, there was no mathematician of note in the whole Christian world," one historian wrote. There were no more advances in geometry until the early 1600s.

Thanks to ancient technology, mathematics did not die, however. Eventually, scholars rediscovered ancient math books and scrolls stored in libraries and monasteries. The rediscovery of advanced mathematics from ancient times jump-started modern mathematics.

Glossary

algebra—a branch of mathematics that deals with quantities expressed in symbols

axiom—a statement accepted as true that serves as a basis for further arguments or theorems

base-10—involving a numbering system in which place values increase in powers of 10

geometry—a branch of mathematics that deals with the measurement, properties, and relationships of points, lines, angles, surfaces, and solids

lunar year—a time period based on the phases of the Moon, lasting a total of 354 days. The lunar year was divided into 12 months of 29 or 30 days each.

mathematics—the science of numbers

numeral—a symbol used to represent a number

pi—the ratio of the circumference of a circle to its diameter; defined by mathematicians as 22/7, or approximately 3.1416

place-value system—a number system in which numerals hold different values depending on their placement

prime number—a number that can be divided evenly only by 1 and itself

right angle—an angle measuring 90 degrees

scientific notation—a system in which numbers are expressed as a number between 1 and 10 multiplied by a power of 10. For example, 29,300 is written 2.93×10^4.

solar year—the time it takes Earth to make a full revolution around the Sun: 365 days, 5 hours, 48 minutes, and 46 seconds

surveying—using mathematics to measure the size and elevation of fields, mountains, valleys, and other physical formations

theorem—a statement in mathematics that has been proved or is to be proved

trigonometry—the study of the properties of triangles

Selected Bibliography

Adkins, Lesley, and Roy A. Adkins. *Handbook to Life in Ancient Rome.* New York: Facts on File, 1994.

Asimov, Isaac. *Asimov on Numbers.* Garden City, NY: Doubleday, 1977.

Benson, Elizabeth P. *The Maya World.* New York: Thomas Y. Crowell Company, 1977.

Fauvel, John, and Jeremy Gray, eds. *The History of Mathematics: A Reader.* New York: Macmillan, 1987.

Fleet, Simon. *Clocks.* London: Octopus Books Limited, 1972.

Grimal, Nicolas. *A History of Ancient Egypt.* Cambridge, MA: Blackwell Publishers, 1994.

Heilbron, J. L. *Geometry Civilized: History, Culture and Technique.* New York: Oxford University Press, 1998.

Hodges, Henry. *Technology in the Ancient World.* New York: Alfred A. Knopf, 1977.

Hollingdale, Stuart. *Makers of Mathematics.* New York: Penguin Books, 1991.

Ingpen, Robert, and Philip Wilkinson. *Encyclopedia of Ideas That Changed the World: The Greatest Discoveries and Inventions of Human History.* New York: Penguin Books, 1993.

James, Peter, and Nick Thorpe. *Ancient Inventions.* New York: Ballantine Books, 1994.

Novikov, Igor D. *The River of Time.* Cambridge: Cambridge University Press, 1998.

Robinson, Andrew. *The Story of Writing.* New York: Thames and Hudson, 1995.

Saggs, H. W. F. *Civilization Before Greece and Rome.* New Haven, CT: Yale University Press, 1989.

Internet Sources

Math Forum, The. n.d. <http://forum.swarthmore.edu> (11 January 2000).

O'Connor, John J., and Edmund F. Robertson. "History of Mathematics." December 1999. <http://www-history.mcs.st-andrews.ac.uk/history/index.html> (11 January 2000).

INDEX

abacus, 45, 47, 49–52
Alexandria, 63, 66, 70–72
Americas: base-20 system, 56–57; holy numbers, 58; Maya calendars, 58–59; reverence for mathematicians, 59–60; zero, Maya use of, 57
Antikythera mechanism, 74–75
Archimedes, 68–70
astronomy, mathematical, 25–26, 71

beam scales, 23–24

calendars: Egyptian, 40; Maya, 58–59; Roman; 81–83
China: abacus, 49–52; counting boards and sticks, 49; stick numerals, 48–49
clepsydra. *See* water clocks
computers, 49–51, 74–75
counting boards and sticks, 49

eclipses, 25–26
Egypt: calendars, 40; geometry, invention of, 36–38; Golenishchev Papyrus, 30, 34; Great Pyramid at Giza, 35; hieroglyphics, 30; knot measurements, 35; mathematics, 33–34; Nilometer, 36; Rhind Papyrus, 30–32, 34; surveying techniques, 35; textbooks, 30–33; timekeeping, 39–40; weight measurement, 38–39
Euclid, 66–68

Greece: Antikythera mechanism, 73–75; Archimedes, 69–70; Diophantus and algebra, 70; Euclid and geometry, 66–68; Greek numerals, 65; Hypatia, 71–72; Pythagoras, 64–66; time measurement, 72–74, 82–83; Tower of the Winds, 73–74; Zeno's Paradoxes, 75–76

Hero of Alexandria, 75
Herodotus, 37–38
hieroglyphics: Egyptian, 30; Maya, 56
Hypatia, 71–72

India: Hinduism and mathematics, 46–47; numeral system, 44–45; place-value system, 45–46; trigonometry, 47

Julian calendar, 82–83

knotted cords for recording numbers, 14–15
Ktesibios of Alexandria, 72–73

mapmaking, 21
mathematics: algebra, 70; calculus, 70; Chinese, 47–52; division, 33–34; Egyptian picture-numbers, 30; geometry, 37–38, 66–67, 83; Indian place-value system, 45–46; Maya, 56–60; Greek, 63–76; multiplication, 33–34, rediscovery of advanced

mathematics from ancient times, 83; trigonometry, 47, 71
mathematics, applied, 63, 79–80
mathematics, theoretical, 63
measurement: land (Egyptian), 35, 37–38; land (Middle Eastern), 20; using body parts, 15–16; weight, 22–24
Middle East: discovery of pi, 26; mapmaking, 21; measuring time, 24–25; measuring weight, 22–24; numbering system based on 60, 25; surveying techniques, 19–20; trade and sales, 21–22

numbering systems: base-10, 55–56; base-20, 56–58
numerals: Arabic, 45; Greek, 65; Roman, 80–81

odometers, 75, 81

pi, 26, 69–70
place-value systems, 45–46, 57–58
Pythagoras, 64–65

Roman Empire, 79–80, 83
Rome: calendars, 81–83; development of first odometer, 81; fall of Roman Empire, effect on mathematics in Europe, 83; Roman numerals, 80–81

scales, 22–23
scientific notation, 70
shadow clocks, 39
stick numerals, 48–49
Stone Age: knotted cords for recording numbers, 14–15; measuring with body parts, 15–16; tally sticks, 14
sundials, 24–25, 39, 73
surveying techniques: Egyptian, 35–36; Middle Eastern, 20

tally sticks, 14
textbooks, mathematical, 30–31
time measurement, 24–25, 39–40, 71–74
Tower of the Winds, 73–74

units of measure: cubit, 16, 21–22; foot, 16, 22; hand, 16, 22; inch, 16

water clocks (clepsydra), 39–40, 72–73
weight measurement, 22–24, 38–39
weight standards, 23–24

zero, 45–46, 55, 57

Note: There are alternate spellings for some of the names mentioned in this book. Here are three examples:
Asoka or Ashoka (India)
Ktesibios or Ctesibius (Greece)
Elea or Velia (Greece)

About the Authors

Michael Woods is an award-winning science and medical writer with the Washington bureau of the *Toledo Blade* and the *Pittsburgh Post Gazette.* His articles and weekly health column, "The Medical Journal," appear in newspapers around the United States. Born in Dunkirk, New York, Mr. Woods developed a love for science and writing in childhood and studied both topics in school. His many awards include an honorary doctorate degree for helping to develop the profession of science writing. His previous work includes a children's book on Antarctica, where he has traveled on three expeditions.

Mary B. Woods is an elementary school librarian in the Fairfax County, Virginia, public school system. Born in New Rochelle, New York, Mrs. Woods studied history in college and later received a master's degree in library science. She is coauthor of a children's book on new discoveries about the ancient Maya civilization.

Photo Acknowledgments: National Museum, Karachi/Art Archive, p. 1 (det.); Trustees of the British Museum, pp. 2–3; The Granger Collection, New York, pp. 11 (det.), 50; Archivo Iconografico, S.A./CORBIS, pp. 12–13; Australian Picture Library/Art Archive, p. 15; Gianni Dagli Orti/CORBIS, pp. 17 (det), 20; Art Archive, pp. 18–19, 31, 37; © Ronald Sheridan/Ancient Art and Architecture Collection, Ltd., pp. 22, 23, 28–29, 38, 56, 72, 82; Roger Wood/CORBIS, p. 27 (det.); Photo: AKG London/VISIOARS, p. 32; © Liu Liqun/ChinaStock, pp. 41(det), 48; (c) Paolo Koch/Photo Researchers, pp. 42–43; © Dr. D. Rajgor/Dinodia Picture Agency, p. 44 (both); Bettmann/CORBIS, p. 51; Private Art Collection/Bridgeman Art Library, London/New York, pp. 53 (det.), 54–55, 68; © B. Norman/Ancient Art and Architecture Collection, Ltd., p. 59; Ashmolean Museum, Oxford, U.K./Bridgeman Art Library, London/New York, p. 61 (det.); Museo Archaeologico Nazionale, Naples, Italy/Roger-Viollet, Paris/Bridgeman Art Library, London/New York, pp. 62–63; Capitoline Museum, Rome/Art Archive, p. 64; The Bodleian Library, the Universityof Oxford, Ms. D'Orville 301, f. 325v, p. 67; Sandro Vannini/CORBIS, p. 74 (det.); Photo: AKG London/Gilles Mermet, pp. 77, 78–79.

Front cover: (© D. Donne Bryant/DDB Stock Photo (left); National Museum, Karachi/Art Archive (bottom); Independent Picture Service (background).

Back cover: Independent Picture Service.

Maya calendar from Tikal, Guatemala

century A.D. and were more accurate than any other calendars in the ancient world.

She Computed

Mathematicians held an honored spot in Maya society. They helped keep the calendar, and they predicted the movements of celestial bodies. Mathematicians also made business calculations, computing prices of goods and land, for instance.

Mathematicians were shown in Maya picture-writing by a special symbol that included a scroll with numbers. The first mathematician identified in Maya picture-writing was a woman. We don't know her name or anything about her. But she must have been very important.